our
ONE
ocean

a story in time

Louise Marsh
Cameron Sesto
Victor Young

Illustrations by Alice Johnson

Teaching One Ocean

Dedicated to Deborah Cramer, author of

Smithsonian Ocean: Our Water Our World

and to the children who love

our one ocean.

Let us imagine going back in time billions of years. What did the ocean look like? When did the first fish crawl ashore from shallow waters and venture onto land? Where does more than half of the oxygen we breathe come from? These are just some of questions **Our One Ocean** will answer.

This story is a history of the ocean, a visual journey through time. Beginning with the big bang, our story moves through billions of years to the formation of the solar system, to the beginning of life in the sea, to the evolution of fish and land animals, ending with the arrival of humans.

The narrative will catch the imagination of young children, as they take their first steps into the magical and spectacular world of ocean science. Some children will be inspired to continue the journey with wonder and curiosity, perhaps becoming future astronomers, paleontologists, or marine biologists.

Everyone who reads this story will come to celebrate the ocean and understand the precious life it has given us, a life that is still ever evolving.

Long, long ago,

the longest time
we know,
the Universe
began . . .

13.7 billion years ago

Time, space, and matter, the building blocks of our universe, began with a cosmic event.

with a

sudden

expansion of space
and energy.

Scientists refer to this expansion as the "Big Bang."

Scientists call our Earth the "Goldilocks Planet." It is not too cold or too hot.

Long, long ago . . .

planets swirling in space revolved around a fiery star, **our sun**.

The sun is the center of our **solar system**.

Earth, our home, is the third planet from the sun.

It is just the right distance from the sun for water to exist as a liquid, essential for most life.

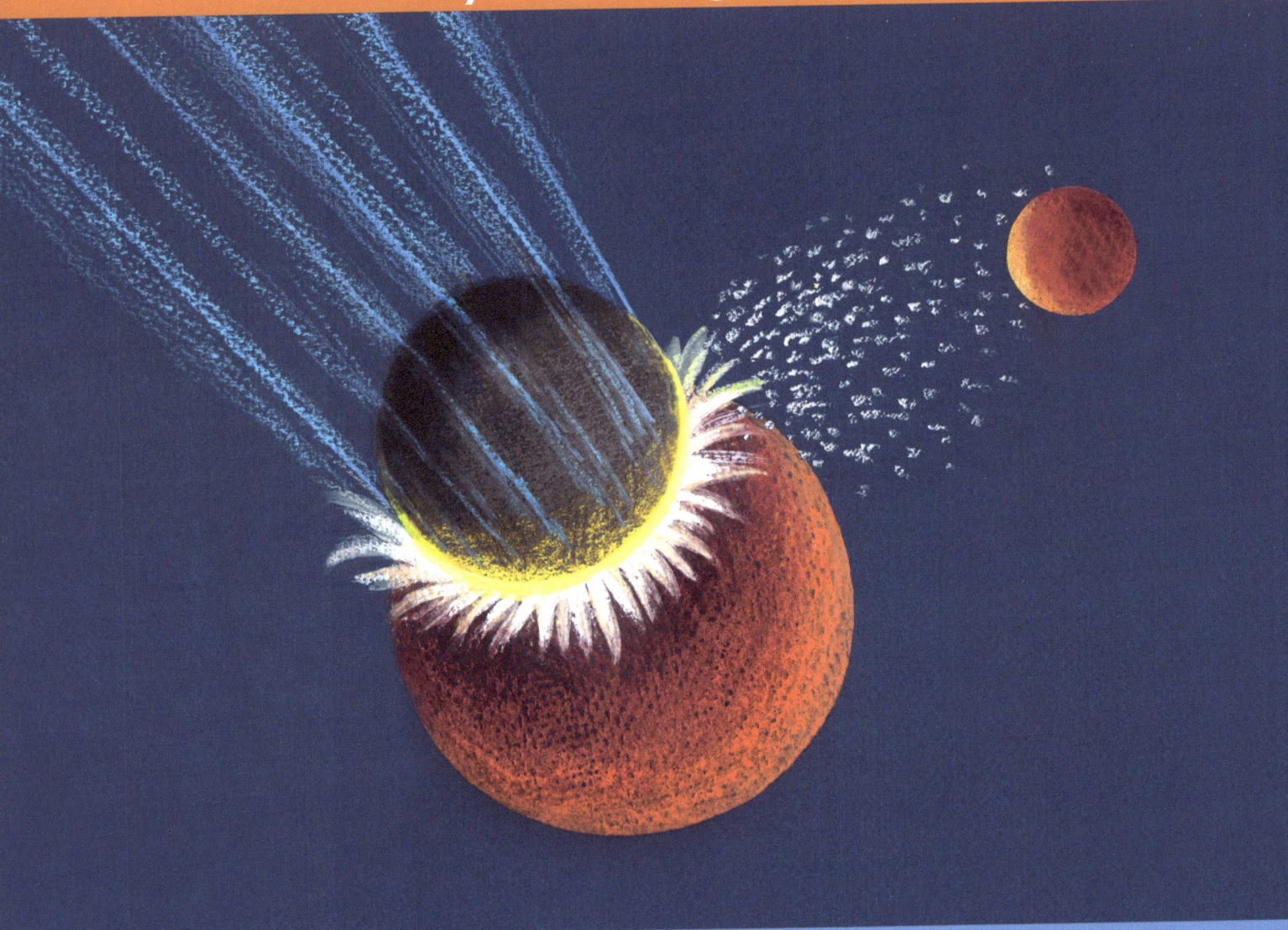

Billions of years ago the moon was much closer to earth than it is today. Then, the pull of the moon's gravity was so strong, it caused huge, massive tides, a thousand times higher than today.

Long, long ago . . .

a huge asteroid crashed into **Earth**.

Chunks of molten, hot Earth blasted into space.

The spinning pieces of Earth **collided**, clumped together and formed our **moon**.

Our moon still causes the rise and fall of the ocean waters, the ebb and flow of the tides.

Of the inner planets of our solar system, Mercury, Venus, Earth, and Mars, only Earth has an ocean. There are different theories on how our ocean formed. One theory is that steam from the volcanoes condensed, causing it to rain for thousands of years.

Long, long ago . . .

volcanoes **exploded** all over the young planet.

Hot steam from the volcanoes cooled, **condensed**, and fell to the earth as **rain**.

Another theory is that comets and meteorites bombarded the young planet. Comets carrying ice and water crashed into the Earth providing enough water to form our ocean.

If you were to travel back in time billions of years, you would not recognize the ocean we know today. The ocean was hot and full of toxic chemicals. No plants and animals could live there.

Long, long ago . . .

our **ocean** was boiling **hot**.

Its color may have been
olive green.

It was an ocean that **no one**
would want to swim in—
not even a fish!

Even the air in the atmosphere was very different from the air
we breathe today. Poison gases were erupting from volcanoes.
There was no oxygen in the air to breathe.

3.4 billion years ago

First life began on the dark ocean floor. Microscopic cells formed by using heat and nutrients from gases boiling out of the deep sea vents. These single cells are called prokaryotes, which means "first seed."

Long, long ago . . .

first life began in the ocean
as very tiny cells
down,
down,
near the **steamy hot**
deep sea vents.

Today, we can only see procaryotes through a microscope, yet they are essential to all life. They inhabit every part of the earth. They include the primitive form of cells we call bacteria.

2.7 billion years ago

Plant-like cells made their own food through a process called photosynthesis. Photosynthesis means "putting together with light." Plant-like organisms captured energy from sunlight, took in carbon dioxide, and released oxygen into the atmosphere.

Long, long ago . . .

golden **sunlight** warmed the surface of the ocean.

Capturing the energy from the sun and nutrients from the water, plant-like cells made their own **food**!

Today photosynthesis still happens in the ocean. All living creatures, including us, depend on marine plants for more than half the oxygen we breathe.

1 billion years ago

The word plankton means "wanderer" or "drifter." Plankton is made up of zillions of organisms that float freely on the surface of ocean waters, carried along by waves and currents. The plant-like organisms are called phytoplankton.

Long, long ago . . .

the single cells **divided**
and multiplied.

Some of them became multi-cellular
plants and animals of many shapes,
floating in our ocean.

They are called **plankton**.

The tiny animals are called zooplankton. Plankton is the first
link in the marine food chain. Today, fish, mussels, birds,
and even the largest animals on earth, the baleen whales,
feed on plankton.

521 million years ago

Over time, ocean animals adapted ways to survive in their environment. A hard shell or exoskeleton protected them against predators. Jointed legs allowed them to move freely along the ocean floor in search of food.

Long, long ago . . .

some animals developed
hard shells that protected
their **soft bodies**.

These were snails, trilobites,
cephalopods, and
horseshoe crabs.

Today horseshoe crabs can be seen on our beaches burrowing
in the sand to lay eggs. They have lived on the planet for
millions of years, and were even here before the dinosaurs.

400 million years ago

Fish were one the first creatures to have a backbone, which supported muscles that allowed the fish to be larger, stronger swimmers. Fierce fish, like the Dunkleosteus, had thick fins and heavy bones. Its head was covered in tough armored plates.

Long, long ago . . .

ancient fish had strong **backbones** and big teeth.

With **fins** and **tails** they swam in the deep blue ocean.

A green jawless fish resembled a giant tadpole. It had tough plates lining its flat head, and eyes on either side of its skull. It scooped up its food from the mud.

When insects and plants moved onto land, they needed water to survive. Plants developed a root system to collect water from the soil. Insects developed an external skeleton to conserve water within their bodies.

Long, long ago . . .

plants began to grow on **land**.

Some **insects** crawled about on the swampy forest floor.

Others flew through the air on **enormous wings**.

Insects were the first creatures to master flight. The dragonfly had a wingspan of up to two feet or more. It could fly up to the top of the horsetail plant, which was fifty feet high.

375 million years ago

Scientists were amazed when they discovered the fossil of Tiktaalik in 2004. It revealed a "walking fish" that could swim and move onto land. It had a flexible neck that could move up and down and wrist-like bones in its limbs to support its weight.

Long, long ago . . .

many millions of years ago,
a fish crawled out of shallow
waters and onto the shore.

It **breathed air** with its lungs!

When scientists found the fossil of
the fish, they named it **"Tiktaalik."**

The name Tiktaalik is an Inuit word meaning "large shallow
water fish." Tiktaalik is the link between ancient fish and all
four limbed, land animals, the Tetrapods.

360 million years ago

The word dinosaur means "gigantic lizard." They were the reptiles that evolved from amphibians. Amphibians laid soft eggs in water. Dinosaurs laid water-tight eggs with shells that did not dry out on land. They then became land dwelling animals.

Long, long ago . . .

the **dinosaurs** became the rulers
of the planet.

They roamed the air, the land,
and the sea for **millions** of years.

Near the end of the Dinosaur Age, a new animal species
evolved: an agile mouse-like mammal with a furry coat. It
could hunt at night and eat small insects and plants. It survived
after the extinction of the dinosaurs.

150 million years ago

The earliest known bird was a tree-climbing, two-legged dinosaur.
It had feathers, teeth, claws on its wings, and a long, bony tail. It
could walk and perch like a bird.

And then . . .

the **flying** dinosaurs arrived.

With **feathered** wings they could soar above the land and the sea.

Chickens, seagulls, and even robins are surviving relatives of the dinosaurs. These flying creatures developed strong, hollow bones, which made them lighter than land animals.

55.8 million years ago

Marine mammals are very much like us. They share similar characteristics of all land mammals. They are warm blooded, nurse their young, breathe with lungs, and have fur or hair on their bodies.

And still
long, long ago . . .

some **mammals** ventured off the land and returned to live in the ocean.

These are the **whales**, porpoises, sea lions, and the **dolphins** that we see swimming in our ocean today.

Whales evolved from a land mammal. It was a dog–like creature that had a long tail and fed by the water's edge. Its cousin, the hippopotamus, shares a common ancestor with the whale.

Our ocean influences every aspect of modern human life. It puts food on our plates and sustains us with water. It provides us with more than half of all the earth's oxygen. It moderates our climate and weather patterns.

And then **not** so very **long ago**,

there were people living on Earth.

We could **talk** to one another.

We built boats and bridges to cross the ocean.

We **fished** for food.

The ocean helps us transport goods all over the world, connecting people and cultures on every continent. For everyone, our ocean is a source of inspiration, recreation, and enjoyment.

We **studied** the ocean.

We came to understand that
we are part of a great
ocean story, where we must live
in harmony with **all life**.

Life that once began **long, long ago**
as very tiny cells in
our one ocean.

The Rise of Humans

Our One Ocean — an ever changing story in time — is written in the layers of rocks on land and the sediment beneath the sea, where scientists discover fossils. This fossil record tells us about our past: what ancient organisms looked like, how they behaved, and where they lived. By studying fossils scientists can mark the passage of time and understand how the ocean has changed our planet.

Compared to the vast expanse of cosmic and geologic time – over 13.7 billion years – humans have lived on the planet for one brief moment — just a blink of an eye!

Our ocean has supported an amazing diversity of plants and animals: the microscopic single cells, oxygen-producing plankton, animals with shells, the armored swimmers, flying insects, the great dinosaurs, and finally all mammals, including humans.

From recent DNA we know that every living thing on Earth is connected

The Anthropocene

to the first cells that formed deep in the ocean waters. Scientists ranging from geologists to biologists also know that we could be entering a new epoch, the Anthropocene, a time defined by man's impact on earth's fragile resources.

Humans live on dry land, yet we are the one species that has the tools and written language to study the ocean. In years to come we will continue to explore the ocean's frontiers. But from a humble human perspective, may we always wonder at its complexity, its beauty, and its magnificent past.

Authors Note:

Though our one ocean is billions of years old, we are just beginning to understand its importance and complexity. We now know the water covering the Earth's surface is not divided into five oceans and seven seas. There is only one ocean. It is a single body of water that is in constant motion, circulating around all the continents, from the tropics to the arctic poles. Without the ocean's circulating system to regulate Earth's climate, the surface of our planet could freeze at night and be too hot for most life to exist during the day. New discoveries about our one ocean increase daily and with it, the knowledge of how humans influence the ocean's health.

The authors, a science teacher, an art teacher, and a learning specialist from Stoneridge Montessori School, became passionate about improving ocean studies for our students. We imagined an ocean curriculum that could synthesize the current research and tie it directly to other academic subjects in the classroom.

Guided by the *Principles of Ocean Literacy*, we built an ocean timeline curriculum. To illustrate the abstract concept of evolutionary time, we created a real time line: a fifty-foot blue cotton strip painted with colored stripes. The timeline marks key concepts and eras of the ocean's history. Students can walk the length of the line and visualize evolutionary time as they study the ocean.

Our One Ocean is a companion book to the timeline, an introduction to the very centerpiece of our curriculum.

For you to continue the journey through the ocean's history, we invite you to visit our website: **teachingoneocean.com**. There you will find photos of students using the timeline, a teacher's guide you can order, art projects, a bibliography, and lesson plans. To learn about the essential *Principles of Ocean Literacy*, visit:

http://oceanliteracy.wp2.coexploration.org/

We hope you will immerse yourself in a study of the ocean as we have. When you stand on the shores of this critically important topic, you will feel how closely we are connected to all life, life that continues to emerge from our one ocean.

Louise S. Marsh   Victor Young   Cameron Gest

Acknowledgements:

The authors wish to thank teachers Diane Sullivan, Linda Seeley, Margaret Henry, Martha Nikas, and the students of **Stoneridge Children's Montessori School** and **North Shore Montessori School** for their participation in our presentations and art projects. For their generous support and suggestions, we would like to thank: Perrin Chick of the **Seacoast Science Center** in Rye, NH, and Mary Kay Taylor and Curtis Sarkin from **Maritime Gloucester**. We are also very grateful to Tatty Bent and Linette Marsh for their thoughtful guidance.

and special thanks to Tiktaalik for coming ashore!

teachingoneoncean.com

www.ingramcontent.com/pod-product-compliance
Lightning Source LLC
Chambersburg PA
CBHW040749200526
45159CB00025B/1821